LILIE GARDENING HORTICULTURISTS GUIDE FROM CULTIVATION TILL COMMMERCIAL SUCCESS

Handbook for Cultivation, Marketing, and Expert Strategies, Techniques, and Tips for Thriving in the Business World

MANUEL SHELTON

Copyright © 2024 by Manuel Shelton

All rights reserved. No part of this publication may be reproduced, distributed, or transmitted in any form or by any means, including photocopying, recording, or other electronic or mechanical methods, without the prior written permission of the publisher, except in the case of brief quotations embodied in critical reviews and certain other noncommercial uses permitted by copyright law.

Disclaimer

The views expressed in this book are solely those of the author and do not necessarily reflect the views of any company, organization, or individual.

The author is not engaged in any endorsement deals or partnerships with any entities mentioned in this book. Any references to products, services, or organizations are for informational purposes only and do not constitute endorsement.

Readers are encouraged to conduct their own research and exercise their own judgment before making any decisions based on the information provided in this book

Contents

CHAPTER ONE 11
  SELECTING THE RIGHT LILIE TYPES................... 11
    How To Plant Lilie Bulbs 11
    Knowing About Lilie Plants 13
    Organizing Your Garden Area 15

CHAPTER TWO 17
  GETTING THE SOIL READY 17
    Essential Tools And Equipment...................... 18
    Appropriate Methods For Soil Preparation 20
    Methodical Planting Procedure 22

CHAPTER THREE 25
  HYGIENE AND PROTECTION 25
    Watering Routines And Methods..................... 25
    Techniques For Fertilization For Optimal Growth. 27
    Common Diseases And Pests: Preventing And Handling Them 29

CHAPTER FOUR 33
  LILY PLANTS: PRUNING AND TRAINING............... 33
    Pruning Is Essential For Healthier Growth 33
    Methods For Pruning Plants With Lilies............. 34

Developing Lily Plants For Improved Shape And Output 36

CHAPTER FIVE 39

HARVESTING FLOWERS LILI 39

Indications That It's Time To Harvest 39

Methods Of Harvesting To Maintain The Quality Of Flowers 40

Tips For Post-Harvest Handling And Storing 41

CHAPTER SIX 43

SELLING AND MARKETING FLOWERS...................43

Finding Lilie Flowers' Possible Markets 43

Advice On Packaging And Presentation 45

Methods For Setting And Marketing Prices For Lilie Flowers 47

CHAPTER SEVEN 49

AGRROPRESTING YOUR SELF-GARDENING COMPANY 49

Increasing The Size Of Your Garden 49

Boosting Output With Effective Techniques 51

Choosing And Overseeing More Assistance.........53

CHAPTER EIGHT 57

FAQS AND REGULAR QUESTIONS 57
What Are The Best Conditions For Lilie Plants To Grow In? 57
How Do I Handle Common Illnesses And Pests? .59
Which Lilie Types Are Most Popular In The Marketplace? 62

CHAPTER NINE 65
INVESTIGATING AND RESOLVING ISSUES 65
Solving Frequently Occurring Problems In Lilie Gardening 65
Techniques For Solving Problems In The Garden 68
Resources For Additional Support And Education 70

CONCERNING THIS BOOK

"Lilie Gardening: Cultivating Beauty and Business" takes readers on a fascinating tour into the world of Lilie plants and provides both novice and expert gardeners with a thorough guide on caring for these beautiful blooms. This book is a true goldmine of information, ranging from the lush pages explaining Lilie plants to the busy marketplace promoting and selling beautiful blossoms.

The voyage commences with a perceptive investigation of Lilie plants, exploring their distinct attributes and necessities. With the necessary equipment and information, readers can easily create the perfect garden area and start their gardening journey with Lilie.

The most important aspect is soil preparation, with thorough talks about soil types and planting methods guaranteeing a strong base for abundant blooms.

The chapters on watering and maintenance are crucial since they reveal to readers how to provide their Lilie plants with the ideal amount of nutrients and moisture.

Gardeners may perfectly care for their blooms by using pest management techniques, fertilization plans, and watering regimens. Their abilities are further developed by pruning and training methods, which improve the beauty and health of their Lilie plants.

The focus shifts to gathering and selling Lilie flowers as the garden blooms. Readers are given the means to deliver perfect petals to eager marketplaces, from spotting ripe blooms to learning post-harvest care. Lilie's industry success can be attributed to the business-savvy approach taken in packaging, pricing, and sales techniques.

The book provides ideas for scaling up operations for people who are prepared to make their gardening endeavors even more ambitious.

Readers are taken through the complexities of building a thriving Lilie business, whether it be through increasing garden space, maximizing efficiency, or managing additional support.

Amid the abundance of information, there is a guiding light in the shape of FAQs, troubleshooting manuals, and other resources. Common issues are confidently addressed, enabling readers to easily pass any obstacles.

Beyond only teaching, "Lilie Gardening" provides a life-changing experience that improves gardens and people.

Throughout the wonderful voyage of lily cultivation and trade, this book is a devoted companion, from the first sensitive sprout to the last gorgeous blossom.

Settlements For Lilie Gardening

Choosing the Ideal Location

Choosing the ideal location for your garden is the first step in starting a Lilie garden. As lily plants prefer bright, sunny spots, pick a spot in your yard that gets at least six hours of direct sunlight per day. Furthermore, make sure the soil in the location you've selected drains properly to avoid waterlogging, which can cause root rot and other problems.

Getting the Soil Ready

It's time to get the dirt ready for your Lilie garden after you've selected the perfect spot. To start, clear the space of any weeds or rubbish so that your plants have a fresh start. Next, use a garden fork or tiller to loosen the soil until it is at least 12 inches deep. This will facilitate better drainage and make it easier for the roots of your Lilie plants to pierce the soil.

CHAPTER ONE

SELECTING THE RIGHT LILIE TYPES

Take your time in selecting the Lilie kinds that are most suited to your growth conditions and preferences before you begin planting. Because lily plants are available in a broad variety of hues, sizes, and bloom durations, you should take your environment, soil type, and preferred aesthetic into account when choosing your plants.

How To Plant Lilie Bulbs

Now that you've chosen your Lilie kinds, the bulbs need to be planted. For every bulb, dig a hole that is roughly three times deeper than the bulb's height, and arrange the bulbs according to the planting guidelines specific to the kinds you have selected. With the pointed end facing up, place each bulb in its designated hole. Cover the bulbs with dirt and carefully press down to eliminate any pockets of air. To help the soil settle and promote root growth, give the freshly planted bulbs plenty of water.

How to Take Care of Lilie Plants

Once your lily bulbs are planted, it's critical to provide them the attention they require to flourish. Water the soil whenever the top inch feels dry to the touch, making sure the soil is continuously damp but not soggy. Additionally, to give your Lilie plants the nutrition they require for strong development and abundant blooming, add a balanced fertilizer once a month during the growing season.

How to Keep Your Lilie Garden Safe

Lastly, take precautions against illnesses and pests that could endanger the health and yield of your Lilie plants in your garden. Look out for common Lilie pests including slugs, snails, and aphids, and take the required action to control them if needed. Additionally, keep an eye out for any symptoms of sickness in your plants, such as fungal or viral infections, and treat them right away to stop the spread of the illness. Your Lilie garden will repay you with lovely blooms year after year if you give it the right care and attention.

Knowing About Lilie Plants

Gardeners all over the world adore lily plants, which belong to the genus Lilium, for their exquisite blossoms and enticing scent. These perennial bulbs are adaptable additions to any garden or landscape because they come in a broad range of hues, forms, and sizes. There is a Lilie variation to fit every taste and growing situation, from the graceful, nodding petals of the Oriental Lilie to the exquisite trumpet-shaped blooms of the Asiatic Lilie.

Lilie Plant Life Cycle

To successfully cultivate and take care of Lilie plants in your yard, you must comprehend their life cycle.

Depending on your climate and the kind you're cultivating, lily bulbs are usually planted in the fall or early spring. Once planted, the bulbs will not sprout again until the spring; instead, they will stay dormant throughout the winter. The plants will keep growing and developing as the weather becomes warmer until they reach their maximum height and blossom in the summer. After flowering, lily plants will go into dormancy again, storing energy for the following growing season while they rest.

Conditions of Growth for Lilie Plants

Creating the ideal growing environment for Lilie plants is essential to their success. These bulbs grow best in locations with lots of sunlight and well-draining soil that has a pH between slightly acidic and neutral. Lilie plants are a great option for novice gardeners because, once established, they require very little care. Your Lilie plants will repay you with years of lovely blooms and delicious scents if you give them the care and attention they need.

Common Lilie Plant Varieties

There are thousands of different Lilie kinds available, and each has special traits and needs for growth. Among many others, popular varieties of lilies include Trumpet lilies, Tiger lilies, Asiatic lilies, and Oriental lilies. There is a Lilie variety to fit your taste and style, whether you want bright, dramatic colors or delicate pastel tones. Try experimenting with several Lilie plant varieties to produce eye-catching displays and gorgeous color combinations in your landscape.

Organizing Your Garden Area

Selecting the Ideal Site

Selecting the ideal spot for your Lilie garden is essential to the success of your plants. For lily plants to flourish, choose a location in your yard that gets at least six hours of direct sunlight per day. Furthermore, make sure that the soil in the area you have selected drains well to avoid standing water, which can cause root rot and other problems.

Creating a Garden Layout Design

It's time to plan your garden layout after deciding on the ideal spot for your Lilie garden. Think about elements like your garden's dimensions and design, as well as the hues and stature of your Lilie plants. Group your plants based on when they bloom and what color they are to create an eye-catching display that will excite the senses all during the growing season.

CHAPTER TWO

GETTING THE SOIL READY

To provide your plants with the greatest growing circumstances possible, you must prepare the soil before planting your Lilie bulbs. To start, clear the space of any weeds or rubbish so that your plants have a fresh start. Next, use a garden fork or tiller to loosen the soil to a depth of at least 12 inches. Then, amend the soil with organic matter, such as old manure or compost, to enhance fertility and drainage.

How to Plant Lilie Bulbs

Now that the soil has been ready, plant your Lilie bulbs. For every bulb, dig a hole that is roughly three times deeper than the bulb's height, and arrange the bulbs according to the planting guidelines specific to the kinds you have selected. With the pointed end facing up, place each bulb in its designated hole.

Cover the bulbs with dirt and carefully press down to eliminate any pockets of air. To help the soil settle and promote root growth, give the freshly planted bulbs plenty of water.

Essential Tools And Equipment

Tiller or Garden Fork

To prepare the soil in your Lilie garden space, you will need a garden fork or tiller. With the help of these instruments, you may loosen the soil to a minimum depth of 12 inches, which will facilitate the roots of your Lilie plants in getting nutrients. Select a garden fork or tiller of superior quality that features robust tines or blades to expedite duties related to soil preparation.

Manual Trowel

A handy instrument that every gardener needs to possess is a hand trowel. Use it to cultivate the soil surrounding your plants, move seedlings, and dig individual planting holes for your Lilie bulbs.

For years of dependable usage in your garden, look for a hand trowel with a durable, rust-resistant blade and an ergonomic grip.

Hose or Watering Can

Your Lilie plants need proper hydration to stay healthy and vibrant, especially in the sweltering summer months. To water your plants straight at the base without uprooting the soil or leaves, use a watering can or hose fitted with a soft spray nozzle. Water your Lilie garden thoroughly and evenly, making sure the soil stays regularly damp but not soggy, to encourage strong roots and a profusion of flowers.

Shears for pruning

Throughout the growth season, pruning shears are essential for keeping your Lilie plants looking and feeling their best. Utilize them to manage the size and spread of your plants, remove sick or damaged leaves, and deadhead spent flowers. For ease of use and precise pruning, look for pruning shears with sturdy, sharp blades and ergonomic handles that fit your hand well.

Appropriate Methods For Soil Preparation

The conditions for good lily development are set by preparing the soil before planting. To guarantee your lilies have the greatest start possible, use these techniques:

Getting Rid of the Space:

To make room for your lilies to develop, clear the planting area of any weeds, rocks, or other debris. Before planting, weeds must be removed since they may compete with lilies for nutrients and moisture.

Amending the Soil:

To increase the soil's texture and fertility, you might need to amend it, depending on the type of soil you have and its pH levels.

Compost, well-rotted manure, or peat moss are examples of organic matter that can be added to the soil to improve its quality and add nutrients that are necessary for lily growth.

Tilling the Soil:

Till the soil to a depth of 8 to 12 inches with a garden fork or tiller. Enhancing soil aeration, drainage, and root penetration, fosters the development of a lily-friendly environment.

Bringing the Soil Level:

After tilling, use a rake to level the soil's surface and make a planting bed. In addition to preventing water from collecting around the roots, which can cause rot, this guarantees even water distribution.

Methodical Planting Procedure

To properly plant your lilies, adhere to these instructions:

When choosing bulbs, look for firm, healthy bulbs that don't show any damage or illness. Larger bulbs usually provide robust, larger blooms.

Excavating Pits:

Planting holes should be dug two or three times deeper than the bulb's height. To give the lilies room to stretch out as they grow, space the holes at least 8 to 12 inches apart.

Fertiliser Application: Follow the manufacturer's recommendations and incorporate a well-balanced fertilizer into the bottom of each planting hole. This promotes robust growth and supplies vital nutrients for early root development.

Setting Up Bulbs:

With the pointed end pointing up, put each bulb in the center of the planting hole. Fill the hole back in with soil, pressing it in a little to eliminate any air spaces surrounding the bulb.

Watering: To help the soil settle and encourage root growth, give the lilies plenty of water after planting. To promote healthy plant development during the growing season, keep the soil evenly moist but not soggy.

Mulching: To retain moisture, control weeds, and manage soil temperature, cover the base of the lilies with a layer of organic mulch, such as straw, crushed bark, or compost.

Staking (if required): Tall lily cultivars may need to be staked to keep their stems from bending or breaking during severe winds. When it comes time to plant the bulbs, surround them with stakes or plant supports to provide them enough support as they grow.

By preparing the soil and planting your lilies according to these guidelines, you'll provide them with the best possible growing conditions and guarantee that they flourish and yield lovely flowers every year.

CHAPTER THREE

HYGIENE AND PROTECTION

There are several important duties involved in keeping a healthy garden, but the most important ones are upkeep and irrigation. You can be sure that your plants are getting the nutrients and care they need to flourish by following proper maintenance and watering procedures. We'll go over efficient watering plans, methods, and upkeep advice in this guide to keep your garden looking great all year round.

Watering Routines And Methods

Creating a regular watering regimen is essential to your plant's well-being. Water requirements differ among plants, so it's important to know what each species in your garden needs. Deep, infrequent watering is often more beneficial to most plants than regular, shallow watering. In addition to promoting root development, this makes plants more drought-tolerant.

Watering in the early morning or late afternoon reduces evaporative water loss. This is one efficient watering strategy. Furthermore, watering from the base of the plant instead of above lowers the chance of fungal illnesses and guarantees that water reaches the roots, which are where it's most required. If you want to efficiently distribute water straight to the root zone while minimizing water waste, think about utilizing drip irrigation systems or soaker hoses.

Several variables, including plant size, soil type, and weather, affect how much water your plants require. When you notice symptoms of overwatering or underwatering, such as withered foliage or soggy soil, make necessary adjustments to your watering plan. Mulching the area surrounding plants can help hold onto soil moisture and cut down on how often they need to be watered.

Techniques For Fertilization For Optimal Growth

For maximum growth, give your plants the proper nutrients in addition to watering them. Fertilization adds back vital nutrients to the soil that may have been lost over time.

It's crucial to carry out a soil test to ascertain the precise nutritional requirements of your garden before adding fertilizer.

Fertilizers come in different varieties, such as organic and synthetic varieties. Over time, organic fertilizers—such as manure or compost—improve soil fertility and structure by releasing nutrients gradually.

Conversely, synthetic fertilizers supply nutrients more quickly but might need to be applied more frequently.

Carefully follow the directions on the fertilizer container to prevent overfertilization, which can damage plants and increase nutrient runoff. For general plant health, think about using a balanced fertilizer that has equal amounts of potassium, phosphorus, and nitrogen (N-P-K).

As an alternative, focused nutrition can be obtained by using specialized fertilizers made for particular plant species, like flowering or fruiting varieties.

Fertiliser should be applied at the right time and in the right quantity to reduce the chance of nutritional imbalances or toxicity.

Fertilizing in the spring and early summer, when plants are actively growing, guarantees that they have access to the nutrients necessary for rapid growth and development.

After fertilizing, make sure to give plants enough water to allow the nutrients to properly seep into the soil and reach the roots.

Common Diseases And Pests: Preventing And Handling Them

If left unchecked, pests and illnesses may wreak havoc on your garden, so management and prevention are essential to preserving plant health. Promoting a healthy garden ecosystem through appropriate cultural practices, such as crop rotation, companion planting, and maintaining high soil health, is one of the most efficient methods for controlling pests and diseases.

Early detection of common diseases and pests minimizes the chance of widespread outbreaks or infestations and enables timely intervention.

Look for evidence of illnesses such as leaf spots, powdery mildew, or drooping, as well as indicators of insect damage such as chewed leaves, stippling, or wilting.

Strategies for integrated pest management (IPM) incorporate a variety of techniques, such as chemical, mechanical, biological, and cultural procedures, to control pests and diseases.

Cultural methods that help lessen the pressure from pests and diseases include clearing away plant detritus, maintaining good hygiene, and spacing plants properly.

Without the use of chemicals, mechanical methods like handpicking bugs or erecting physical barriers like row coverings can be effective pest management. Biological control is a sustainable and environmentally benign method of managing pest populations by introducing natural predators or parasites to the target population.

When chemical intervention is required, select pesticides that are specifically designed to address the insect or disease issue at hand, and carefully read the label instructions to ensure safe and efficient administration. When feasible, try to use low-toxicity or organic alternatives to reduce damage to wildlife and beneficial insects.

By keeping an eye out for symptoms of pests and illnesses in your garden regularly, you can take action early on and avoid the need for more thorough treatments down the road. You can manage pests and illnesses and encourage a robust and healthy garden ecology by taking proactive steps and keeping a close eye on things.

CHAPTER FOUR

LILY PLANTS: PRUNING AND TRAINING

Lilies are cherished complements to any garden because of their alluring flowers and graceful stature. However, appropriate pruning and training are necessary to guarantee that they reach their maximum potential in terms of health, shape, and productivity. Let's examine the value of pruning for stronger growth, look at several methods for trimming lilies, and talk about training them for the best possible form and yield.

Pruning Is Essential For Healthier Growth

To maintain the general health and vigor of Lily plants, pruning is essential. Removing sick, dead, or damaged leaves and spent blossoms improves the appearance of the plant while also promoting new growth and blooming. Furthermore, trimming lowers the danger of fungal infections and increases photosynthesis by enhancing the plant's ability to breathe and absorb sunlight.

Frequent pruning encourages the lily plant to use its resources more wisely by focusing its energy on developing new leaves, shoots, and blooms. Stronger, more robust plants with a higher potential for blooming are the outcome of this. Additionally, pruning enables you to sculpt the plant to your desired style, making sure it blends in perfectly with the rest of your garden.

Methods For Pruning Plants With Lilies

Using the proper techniques is crucial to getting the best outcomes from Lily plant pruning without damaging the plant. The following are some essential methods to remember:

Deadheading: Eliminate wasted blooms as soon as possible to stop the lily plant from focusing its energy on producing seeds. Just above the closest group of healthy leaves or buds, cut off the faded blossoms with clean, sharp pruners.

Cutting Back Leaves:

Throughout the growth season, trim away any leaves that are yellowing or damaged. To keep the plant looking neat and stop the spread of illness, cut these leaves back to the base of the plant with sterile pruning shears.

Thinning: To promote better airflow and light penetration, thin down your lily plant if it gets too crowded or grows dense leaves. The strongest and healthiest branches should be left intact after extra stems and leaves are removed at ground level.

Stem Pruning: After perennial lily varieties have naturally died back, cut down their stems to ground level in late autumn or early spring. This encourages rapid regrowth in the next season and helps revitalize the plant.

Developing Lily Plants For Improved Shape And Output

Training lily plants entails controlling their growth to maximize flowering yield while achieving a particular shape or form. The following are some practical methods for training lily plants:

Staking: To maintain their stems and keep them from bending or breaking under the weight of blossoms, tall lily varieties like Asiatic or Oriental lilies may need to be staked. Tighten the stems gently with twine or soft ties to the robust pegs you've installed near the base of the plant.

Pruning for Form: Lily plants can be shaped to your liking with regular pruning. Pinch back the tips of immature shoots in the spring to promote the establishment of lateral branches and a bushier growth pattern. As a result, the plant becomes denser, and more compact, and produces several flowering stalks.

Division: Overcrowding of Lily plants over time might result in less vigor and blooms. Every few years, divide the plants in early spring or autumn to encourage healthier growth and revitalize crowded clusters. Dig up the lily bulbs carefully, divide them into smaller clusters, and then replant them at the proper distance apart.

You can make sure that your lilies grow well and provide you with a beautiful display of blooms every season by using these trimming and training procedures in your gardening practice. You can fully appreciate the beauty and grace of lilies in your yard with a little upkeep and care.

CHAPTER FIVE

HARVESTING FLOWERS LILI

Indications That It's Time To Harvest

It's important to know when your lilies are ready to be harvested before you harvest them yourself. The bloom stage is one of the main markers. It is preferable to gather most lily kinds when the flower buds are just starting to open but haven't fully opened. Seek for buds that are beginning to open but haven't quite done so, as well as those that have some color. This phase guarantees a longer vase life for the flowers following harvest. Furthermore, look for any indications of yellowing or injury on the petals. Petals on healthy lily blooms will be strong and flawless.

The color of the flower buds is another indicator of preparedness. Lilies come in a range of hues, from white and yellow to pink and red, depending on the kind. When the buds are a bright, uniform color, that is the best time to pick. Buds that have uneven or poor coloring may not be fully mature and will not store as long after harvesting.

Additionally, notice how the blooms smell. When lilies are ready to be harvested, they usually release a nice scent. The buds could require extra time to grow if they don't smell very strongly.

Methods Of Harvesting To Maintain The Quality Of Flowers

It's critical to pick lily blossoms using the proper methods to guarantee their longevity and quality. Choose a clean, sharp pair of gardening shears or scissors first. Crushing the stems with blunt tools might cause issues with water absorption and hasten wilting. Prepare a clean bucket or vase and fill it with lukewarm water before cutting the flowers. This will aid in hydrating the stems right away following harvesting, maintaining their freshness.

Select buds that are at the perfect level of maturity so that you can harvest the lilies. Just above a leaf node, make a neat, angled incision with the stem held close to the base. Cutting too close to the buds or leaving lengthy stubs might reduce the plant's ability to absorb water and increase the chance of germs developing. After cutting, drop the stems straight into the water-filled container that has been prepared. Handling the flowers lightly is crucial to avoid bruises or damage to the petals.

Tips For Post-Harvest Handling And Storing

For harvested lily blooms to remain of high quality, proper post-harvest handling is essential. To stop bacteria from growing, clip off any leaves that will come into contact with water once the stems have been removed.

For best water absorption, trim the stems at a small angle every few days. To keep the flowers from wilting, keep them out of the direct sun and drafts. If the lilies won't be arranged right away, keep them in a cold, dark spot to prolong their vase life.

Change the vase's water every two to three days to keep it fresh and avoid bacterial development. A flower preservative, which supplies vital nutrients and stops microbiological development, can also assist prolong the life of the lilies in water. To avoid contamination, remove any leaves that will be below the waterline when arranging the lilies in a vase.

Arrange the lilies in a new water-filled vase and keep them in a cool spot away from ripening fruits, which can hasten the aging process of flowers by releasing ethylene gas. Harvested lilies can persist for several days in their brilliant beauty with the right maintenance.

CHAPTER SIX

SELLING AND MARKETING FLOWERS

Selling and promoting Lilie flowers can be a thrilling endeavor, but to successfully reach your target audience, you must take a deliberate strategy.

You can draw clients who value Lilie flowers' grace and scent by showcasing their attractiveness with the appropriate tactics.

Finding Lilie Flowers' Possible Markets

It's critical to take into account both local and internet options when determining possible markets for Lilie flowers. Investigate your area's farmers' markets, garden centers, and florists first.

These companies frequently serve clients who are looking for fresh flowers to present as gifts or to spruce up their houses.

Additionally, think about getting in touch with wedding coordinators, hotels, and venues that might be interested in buying Lilie flowers for special events like weddings, business gatherings, or hotel décor.

You can acquire recurring orders and create a consistent flow of income by building ties with these companies.

Additionally, there is great potential to reach a larger audience through online channels. Think about opening an online store or selling on well-known e-commerce sites like Amazon or Etsy.

By doing this, you may expand your consumer base and reach a wider audience of garden enthusiasts.

Marketing Lilie Flowers can also benefit greatly from the use of social media. To reach potential buyers looking for floral inspiration, create visually appealing posts showing off your flowers in various settings and add relevant hashtags.

Interact with your audience by answering messages and comments right away. You may also think about advertising tailored advertisements to connect with particular audiences.

Advice On Packaging And Presentation

Selling Lilie flowers requires careful consideration of packaging and presentation since they have a big impact on how valuable people think your product is.

When packing your flowers, choose robust boxes or containers to preserve their freshness and safeguard them throughout transportation.

To improve the presentation and provide your consumers with an unforgettable unboxing experience, think about incorporating ornamental elements like ribbon or tissue paper.

Consider color, texture, and arrangement when placing your Lilie flowers on display. Try arranging flowers and greenery in different ways to produce visually arresting arrangements that highlight the inherent beauty of the Lilie flowers.

For an added sense of elegance, think about including complementary items in your display, like elegant pots, baskets, or vases.

High-quality photography is crucial for properly showing your flowers online to increase sales.

Purchase a high-quality camera or work with a professional photographer to get gorgeous photos of your flowers taken from various perspectives.

To make the flowers shine, use natural lighting and muted backgrounds. Then, tweak the images to bring out the details and colors.

Methods For Setting And Marketing Prices For Lilie Flowers

Setting a competitive price for Lilie Flowers is crucial to drawing clients and increasing your revenue. Make sure your prices are reasonable and leave room for a healthy profit margin by researching the costs of comparable flowers both locally and online.

Take into account elements like the flowers' uniqueness, the market's seasonality, and any extra services you provide, like delivery or personalized arrangements.

Providing specials or discounts can also draw clients and promote recurring business. Think of providing exclusive discounts for devoted clients, seasonal promotions, or special offers for large orders. This not only encourages people to buy from you, but it also fosters word-of-mouth recommendations and brand loyalty.

In the floral business, providing exceptional customer service is essential to cultivating a devoted clientele. Respond quickly to questions and requests, and make an effort to go above and beyond for your clients by providing high-quality flowers and rapid delivery.

Invite consumer input and make use of it to keep your goods and services getting better. By giving your clients an outstanding experience, you may win their trust and loyalty, which will guarantee recurring business and long-term success.

CHAPTER SEVEN

AGRROPRESTING YOUR SELF-GARDENING COMPANY

Increasing The Size Of Your Garden

Increasing the size of your garden can be a fun project, but success depends on careful design and execution.

Assessing your current space and figuring out how much extra room you need is one of the initial tasks. When choosing a new place for your extended garden, take accessibility, sunshine exposure, and soil quality into consideration.

After choosing a good location, you must thoroughly prepare the soil before planting. Tilling the ground, adding compost or other organic matter, and taking care of any drainage problems could all be part of this.

Spend some time designing your garden, keeping in mind the kinds of plants you wish to cultivate and how far apart they should be spaced.

Invest in high-quality tools and equipment to facilitate effective management of your enlarged garden. To guarantee proper water distribution, this can involve bigger equipment like wheelbarrows or tillers in addition to extra hoses, sprinklers, or irrigation systems.

To increase production and simplify maintenance, think about adding raised beds or container gardening to your larger garden area. Plants can be grown in regions with poor soil quality or limited space with container gardening, while raised beds can help with soil drainage and weed control.

Lastly, remember to account for the extra time and materials needed to manage your larger garden. Your plants' health and productivity depend on regular irrigation, weeding, and insect management.

Boosting Output With Effective Techniques

To boost output in your Lilie gardening company, you need to combine strategic planning, good time management, and efficient procedures. Start by figuring out where you can cut inefficiencies and streamline your workflow.

Using a crop rotation schedule to make the most of your garden's space and reduce soil erosion is one strategy to increase output.

You can replenish vital nutrients in the soil and help prevent the accumulation of pests and illnesses by rotating your crops every season.

Invest in top-notch tools and equipment to make your gardening task go more smoothly. These could be powered tools like weed eaters or tillers, lightweight wheelbarrows, or ergonomic hand tools. Maintaining and sharpening your tools regularly will also help to make sure they last longer and function as intended.

If you want to control soil temperature, prevent weed growth, and assist preserve soil moisture in your garden beds, think about applying mulch. Straw, wood chips, and shredded leaves are examples of organic mulches that decompose over time and enrich the soil with beneficial organic matter.

Establishing a watering schedule and utilizing soaker hoses or drip irrigation will assist in guaranteeing that your plants get enough moisture with the least amount of water wasted.

To collect rainwater for the garden, think about building a rain barrel or other type of water collection system.

Lastly, to maximize your time in the garden, arrange your tasks according to significance and urgency. To reduce your burden, divide more complex projects into smaller, more doable tasks and assign responsibility when you can.

Choosing And Overseeing More Assistance

You might need to bring on more staff if your Lilie gardening company expands to meet demand.

When recruiting new staff members, seek out people who have a strong interest in gardening together with the qualifications and expertise needed to make a significant contribution to the success of your company.

Clearly define the duties and expectations for every position in your gardening company by creating job descriptions. By doing this, you can make sure that everyone is aware of their responsibilities and what is expected of them.

Make sure new hires receive in-depth training so they are knowledgeable about your gardening techniques, supplies, and machinery. In order to answer any queries or worries they may have and promote a healthy work atmosphere, encourage open communication and feedback.

Put in place mechanisms to monitor worker hours, tasks accomplished, and output to assist you in improving workforce management. Try utilizing applications or scheduling tools to help team members communicate and coordinate more easily.

Establish measurable performance targets and rewards for your staff to inspire them and acknowledge their diligence and hard work. This could be in the form of rewards, bonuses, or chances to progress in your gardening company.

To assist staff members in becoming better and more productive in their positions, regularly assess their performance and offer helpful criticism.

To keep everyone's workplace happy and productive, immediately and fairly handle any issues or concerns.

You can assemble a solid and dependable crew to assist the expansion and prosperity of your Lilie gardening enterprise by adhering to these hiring and management strategies for further assistance.

CHAPTER EIGHT

FAQS AND REGULAR QUESTIONS

What Are The Best Conditions For Lilie Plants To Grow In?

Although lilies are beautiful flowering plants that can grow in many different environments, giving them the right conditions will guarantee that they bloom to their greatest potential. The following information will help you to establish the ideal environment for your lilies:

Sunlight: Since lilies adore the sun, pick a spot in your yard that gets at least 6 to 8 hours of direct sunlight per day. Certain cultivars, such as Martagon lilies, can take some partial shade, though.

Soil: For lilies, well-draining soil is essential. They prefer well-drained, slightly acidic soil over neutral soil since waterlogging can cause root rot. Enhancing soil structure and fertility can be achieved by adding organic matter, such as compost or peat moss.

Watering: Although lilies require frequent irrigation, it's critical to prevent overwatering because wet soil can lead to bulb rot. Water the soil deeply but sparingly, letting the top inch of soil dry out in between applications. Mulching the area surrounding the plants' bases can aid in weed suppression and moisture retention.

Temperature: Lilies grow best in mild climates, with a preference for 60–75°F (15–24°C) during the day and colder evenings. Giving the plants some shade during the hottest part of the day can help them avoid becoming stressed out by excessive heat.

Air Circulation: To avoid fungal infections and encourage healthy growth, there must be adequate air circulation. Lilies shouldn't be planted in congested spaces with restricted airflow. Enhancing airflow around plants can also be achieved by pruning overloaded foliage.

Fertilizing: Since lilies are moderate feeders, they gain from an early spring application of a balanced fertilizer when new growth begins to appear. Steer clear of high-nitrogen fertilizers since they may encourage dense foliage at the price of blooms. Lilies respond well to slow-release fertilizers or liquid fertilizers diluted to half-strength.

Give your lilies the ideal growing environment, and you'll be rewarded with a breathtaking display of vibrant blooms every year.

How Do I Handle Common Illnesses And Pests?

Even though they require little care, lilies are susceptible to pests and illnesses. Here's how to recognize and handle a few typical problems:

Aphids: These tiny, pliable insects gather together on the undersides of leaves and flower buds, where they extract plant sap. Aphids are frequently removed by giving them a powerful water blast from a hose or by treating the affected regions with insecticidal soap. Natural predators like lacewings and ladybirds can aid in the management of aphid populations.

Lily insect: If left unchecked, the vivid red lily insect and its larvae can swiftly defoliate lily plants. Select adult and larval beetles by hand from the plants, then submerge them in a pail of soapy water to kill them. Lily beetle control can also be achieved with neem oil or insecticidal soap.

Grey mold, or botrytis blight, is a type of mold that grows best in chilly, wet environments and leaves brown patches on leaves and blossoms. Reduce the amount of foliage surrounding the plants and steer clear of overhead irrigation to increase air circulation. Any contaminated plant material should be removed and disposed of to stop the disease from spreading.

Fusarium Wilt: This fungal disease causes wilting, yellowing of the leaves, and reduced growth in lilies by attacking their vascular system. Fusarium wilt has no known cure, thus avoidance is essential. The disease can be stopped from spreading by planting lilies in well-draining soil, avoiding overhead irrigation, and eliminating and destroying diseased plants.

Viruses: Stunted growth, diminished flowering, and mottled or twisted leaves can all be caused by the lily mosaic virus and other viral diseases. Since viral infections have no known cure, diseased plants should be removed and destroyed right away to stop the virus from infecting healthy plants. Using clean instruments and maintaining proper hygiene can help stop the spread of viruses.

You can maintain the health and prosperity of your lilies by quickly recognizing and resolving pest and disease problems.

Which Lilie Types Are Most Popular In The Marketplace?

Lilies have a lengthy vase life and spectacular beauty, which makes them attractive flowers in the commercial sector. For commercial use, some of the most popular lily cultivars are listed below:

Oriental Lilies: In the floral business, oriental lilies are highly prized for their big, fragrant blooms and striking hues. 'Stargazer,' 'Casa Blanca,' and 'Sorbonne' are popular varieties for arrangements and bouquets.

Asiatic Lilies: Asiatic lilies are highly valued for their vivid hues and abundant flowering patterns. They are useful for flower design since they are available in a variety of colors, from vibrant primaries to gentle pastels. Top-selling varieties include "Enchantment," "Tiny Sensation," and "Matrix."

Longiflorum flowers, or Easter Lilies: Easter and springtime festivities are often associated with Longiflorum flowers. They are a favorite for both festive displays and religious events because of their gorgeous, trumpet-shaped flowers and delicious aroma.

LA Hybrid Lilies: LA hybrid lilies combine the greatest qualities of both parents through a cross between Asiatic and Longiflorum lilies. They are well-liked for both retail and wholesale markets because of their robust stems, durable blooms, and variety of colors.

Trumpet Lilies:

A gorgeous focal point in bouquets and arrangements, trumpet lilies have a captivating aroma and dramatic, trumpet-shaped blossoms. Beautiful and elegant varieties, such as "African Queen," "Regale," and "Golden Splendour," are highly valued.

Double Oriental Lilies: These lilies have extra petals that give them a rich, luscious look. These striking flowers are a favorite for bridal bouquets and special occasions because they lend sophistication and romanticism to any design.

Commercial farmers and florists can satisfy the market need for premium flowers by providing these well-liked lily varieties, pleasing customers with their beauty and smell.

CHAPTER NINE

INVESTIGATING AND RESOLVING ISSUES

Solving Frequently Occurring Problems In Lilie Gardening

Lilie gardening can be very fulfilling due to its beauty and aroma, but it is not without difficulties. The following are some typical problems you could run into along the route, along with solutions:

Yellowing Leaves: A prevalent problem with Lilie plants is the presence of yellowing leaves. Pest infestations, nutritional shortages, overwatering, and underwatering can all contribute to this. To troubleshoot, measure the moisture content of the soil, modify the amount of irrigation as necessary, and look for pest activity. To supply vital nutrients, think about using a balanced fertilizer as well.

Stunted Growth: Inadequate sunlight, crowded conditions, or poor soil quality may be the cause of your Lilie plants' slower-than-expected growth. Examine the soil's conditions and add organic matter as needed to address this problem. To encourage healthy growth, make sure your Lilie garden gets enough sunlight—at least 6 to 8 hours each day—and prune down any areas that are too crowded.

Wilting or Drooping: Drooping or wilting leaves may be a sign of several issues, including pest damage, root rot, overwatering, or underwatering. Check the moisture content of the soil and modify watering as necessary for troubleshooting.

Make sure the soil is properly drained to avoid soggy patches and look for indications of decay in the roots. Use organic pesticides as soon as possible to treat any pest infestations.

Flower Bud Drop: When flower buds fall off before they open up, it can be rather annoying. Inadequate care, dietary deficits, and environmental stressors can all contribute to this.

Make sure your Lilie garden is in a decent spot with enough sunlight and airflow before trying to troubleshoot. To supply necessary nutrients and preserve ideal growing conditions, fertilize regularly using a balanced fertilizer.

Pest Infestations: Aphids, mites, and slugs are just a few of the pests that lily plants can get in their leaves and blossoms.

Examine your plants frequently for indications of pest activity, then take the necessary precautions, like using neem oil or insecticidal soap, to eliminate the infestation. To help naturally control pest populations, think about luring beneficial insects like ladybirds and lacewings.

Techniques For Solving Problems In The Garden

Like any gardening, Lilie gardening will inevitably provide difficulties. The following problem-solving strategies will assist you in successfully overcoming garden-related obstacles:

Observation and Diagnosis: Pay close attention to your plants and make a diagnosis as the first step in resolving any gardening problem. Look for signs like wilting, yellowing leaves, or odd growth patterns, and investigate potential reasons.

Research and Education: To learn more about Lilie's growing methods and typical issues, make use of resources including gardening books, internet discussion boards, and university extension programs. Knowing what your Lilie plants need will help you make better judgments and address issues more skillfully.

Integrated Pest Management (IPM): By using an integrated strategy, pest populations can be reduced with the least amount of harm to the environment and beneficial insects. This strategy for managing pest infestations combines chemical, biological, and cultural management techniques.

Soil Testing and Amendment: The health and vibrancy of your Lilie garden are greatly dependent on the quality of the soil. You can find nutrient imbalances or deficiencies, pH difficulties, and other soil concerns by doing a soil test. To give your Lilie plants the best possible growing environment, adjust the soil as needed with organic matter, lime, or other soil amendments based on the findings.

Regular Upkeep: The best defense against garden issues is frequent prevention. To maintain the resilience and health of your Lilie garden, follow a routine for watering, fertilizing, trimming, and weeding. You can reduce the likelihood of difficulties by being proactive and aware of your garden's requirements, and you can quickly resolve any issues that do arise.

Resources For Additional Support And Education

For support and continuing education, it's critical to have access to trustworthy resources as you start your Lilie gardening adventure. Here are some worthwhile sites to think about:

Purchasing a few thorough gardening books that address Lilie's maintenance, cultivation, and troubleshooting is a wise investment. Seek for books with helpful information catered to your unique needs and climate, and authored by respected authorities in the subject.

Participate in online forums and communities devoted to gardening to network with other Lilie aficionados and exchange questions and experiences. You can get help and assistance from websites such as GardenWeb, the gardening subreddit on Reddit, and specialized Lilie gardening forums.

University Extension Services: A lot of university extension services provide low-cost or free materials, including workshops, fact sheets, and diagnostic services, for home gardeners. Find out what resources are available in your area by contacting the university extension office in your community.

Garden Centres and Nurseries: Lilie gardeners can benefit greatly from the knowledge and support provided by their local garden centers and nurseries. The employees may provide tailored guidance and suggestions for your garden and are frequently informed about the growing conditions in the area.

Online seminars and Courses: Take into consideration signing up for seminars or online courses that concentrate on Lilie's growing methods and best practices. Experts in the area teach a range of gardening courses on websites such as MasterClass, Coursera, and Udemy.

You may improve your Lilie gardening abilities and conquer any obstacles you encounter by making use of these materials and carrying on with your education and exploration. Keep in mind that gardening is a voyage of learning and development, and every experience—whether positive or negative—adds to your understanding of and pleasure of this fulfilling pastime.